Steam Vehicles Of The Road

Photo Album

Kevin Lomas

Copyright

© Kevin Lomas 2013

All world-wide rights reserved.

No part of this publication may be reproduced in any form without the written permission of the copyright owner.

ISBN - 978-1-291-60047-6

Published by Out Of My Mind

Kevinlomas.net

Kevlms@aol.com

Introduction

My father, as well as being an engineering toolmaker, was a writer, writing articles for magazines. One of his subjects was steam engines, mainly road-use ones: 'Traction (meaning to pull) Engines' (although some later became steam wagons built to carry). As well as gathering some photos from many places, he also took many photos himself between the 1950s and mid-1970s and I have recently found some of both the ones he took and the ones he obtained and here they are in this photo album. I do recall many others, but have no idea what happened to those, or his notes.

Even in that earlier decade the machines were already mostly out of use and photos from the later decades are machines having been restored, or perhaps hoping to be. Some are much older photos of them doing the work they were built for, and some are 'Works' photos taken by the manufactures for promotional purposes.

Steam engines became mounted on to wheeled carts mostly to drag around to farms by horses to drive farming tools by means of belts on flywheels, but no one seems to be 100% sure who first connected the power to the wheels of the base the engines sat on, along with a steering wheel to make them self-propelled although it is said that Thomas Aveling may have done it first in 1859. This is not a history book though, just a photo album.

What is obvious from the photos is that they made an attempt, later, to compete with haulage vehicles that used internal combustion engines to propel them. History proved that they failed. I will do my best to categorise the types.

Without doubt they were wonderful feats of British engineering, heavy engineering, and expensive engineering, nevertheless, compared to the internal combustion engine, they were very inefficient, not least because someone had to shovel coal in to them, and they took a while for the water to heat up. Of course, they were slow, too. They served their purpose well, though, during their around 100 year life-time, all the same, in the great Steam Age, even though some of them did put many farm workers out of a job.

I have attempted to look up the history of the engines in the photos, not having found my late father's notes, and even though some have scanty notes on the back, it has proven to be confusing. As an example, using the name of the engine on just one photo, I find that that engine is possibly still around, however, it's not exactly the same engine now. Also, the 'historics' I have found of Traction Engines on line does not seem entirely correct from what I recall from listening to my Father, as even the few photos I have, plus my own memory of them in use, and what is still in existence today, points out. In view of that, and how many photos there are, this is simply going to be a book of photos and any reader who knows of the engines, the real history of each of them, or even if some are still around (a remarkable number of others have been renovated since the photos were taken, or at least rebuilt from many mixed old and new bits (because one in good running order is very valuable)) feel free to contact me and perhaps if details are found for most of the photos, it may result in another book with far more text in it!

Nevertheless, this should be a worthy book because most if not all of these photos are possibly totally unique.

Kevin Lomas.

Kevin Lomas

Traction Engines

Here, we have what can be called 'true' Traction Engines, ones built to pull things along roads, as well as drive other machinery. Often they also pulled the machinery along roads to the place they were to be used. Note the large flywheels used to temporarily drive belts running to static machinery to drive those.

Kevin Lomas

Steam Vehicles Of The Road

Kevin Lomas

Kevin Lomas

The "Garrett" Compound General Purpose Traction Engines.

Kevin Lomas

Steam Vehicles Of The Road

Steam Vehicles Of The Road

Kevin Lomas

Kevin Lomas

Steam Vehicles Of The Road

Kevin Lomas

Steam Vehicles Of The Road

Steam Vehicles Of The Road

Kevin Lomas

Steam Vehicles Of The Road

Kevin Lomas

Other Traction Engines

Some were also built for very single purposes, almost. Here are three of those purposes.

One was built as the centre and motive power of a fairground ride and no doubt also used to tow the unassembled rides around the country. I will also include a photo of such a ride and also one of a fairground.

Then there were Showmens' Engines. These also towed the rides around the country, but they were also fitted with a large dynamo, over the front of the boiler, to generate electricity for use by the fair. Yes, electricity. These Traction Engines were built to be more ornate, and bristling with coloured light bulbs due to their own dynamos.

Next is the Ploughmans' engine. Longer than a normal Traction Engine, these had large driven winding drums under them, as you can see. Two of these machines would be positioned at opposite ends of large fields and via a system of cables on those drums, would drag an immense plough arrangement across them. Then they would move along a bit and do it again, until the field was completed. It goes without saying they replaced many horses, men, and did the job much faster.

Kevin Lomas

Kevin Lomas

Re-union of Traction Engine Drivers at Nottingham Goose Fair Oct 1951

Kevin Lomas

Steam Vehicles Of The Road

Kevin Lomas

Steam Wagons

As they became more powerful as compared to the weight of them (and roads better surfaced) they were able to build the motive power smaller, however it was more to do with trying to compete with haulage vehicles beginning to utilise internal combustion engines, which you did not have to constantly stoke and even await a decent head of steam before you could even set off. So next we come on to Steam Wagons that began to look more and more like the vehicles they were trying to compete with and towards the vehicles we are use to seeing today.

Kevin Lomas

Steam Vehicles Of The Road

Kevin Lomas

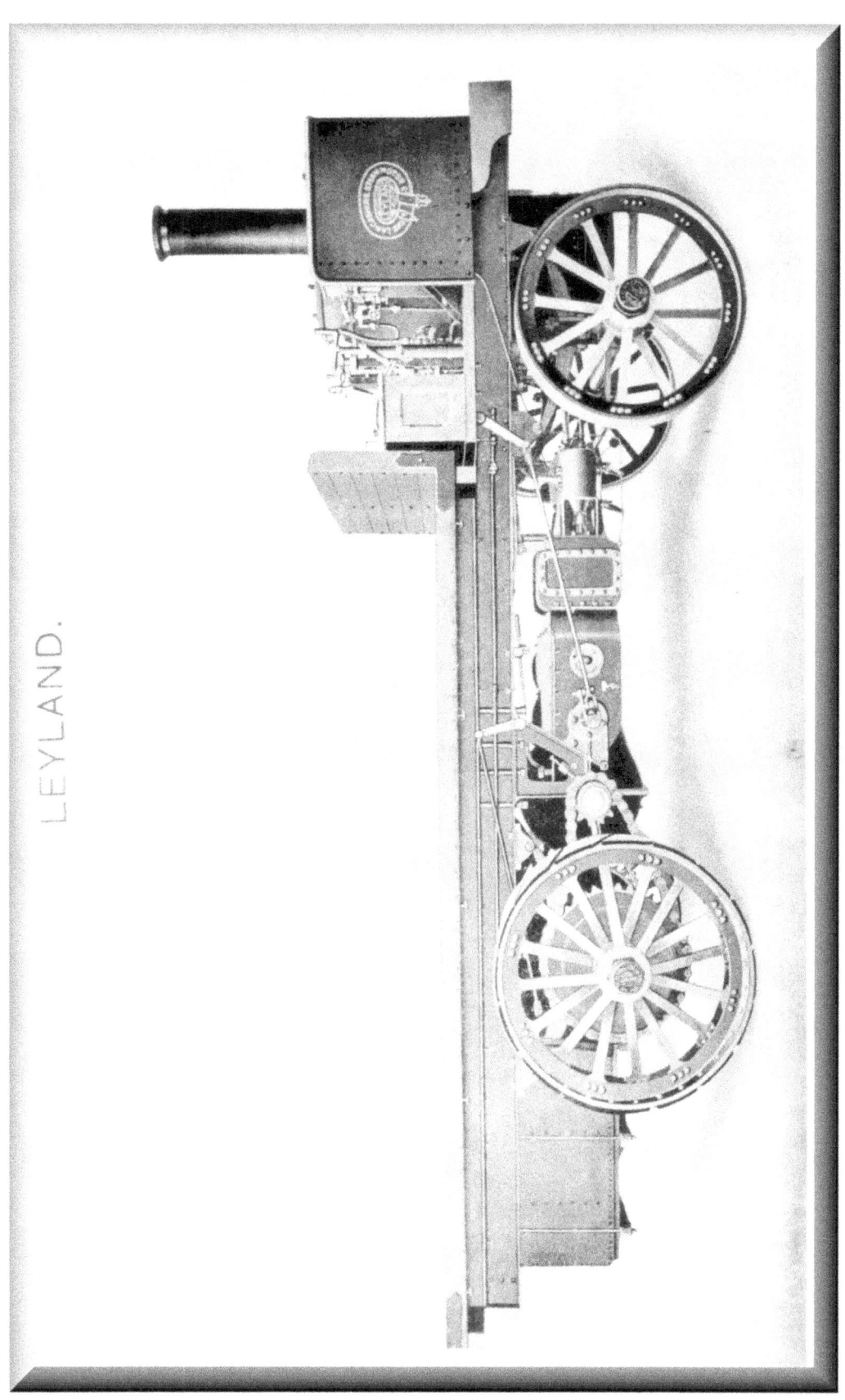

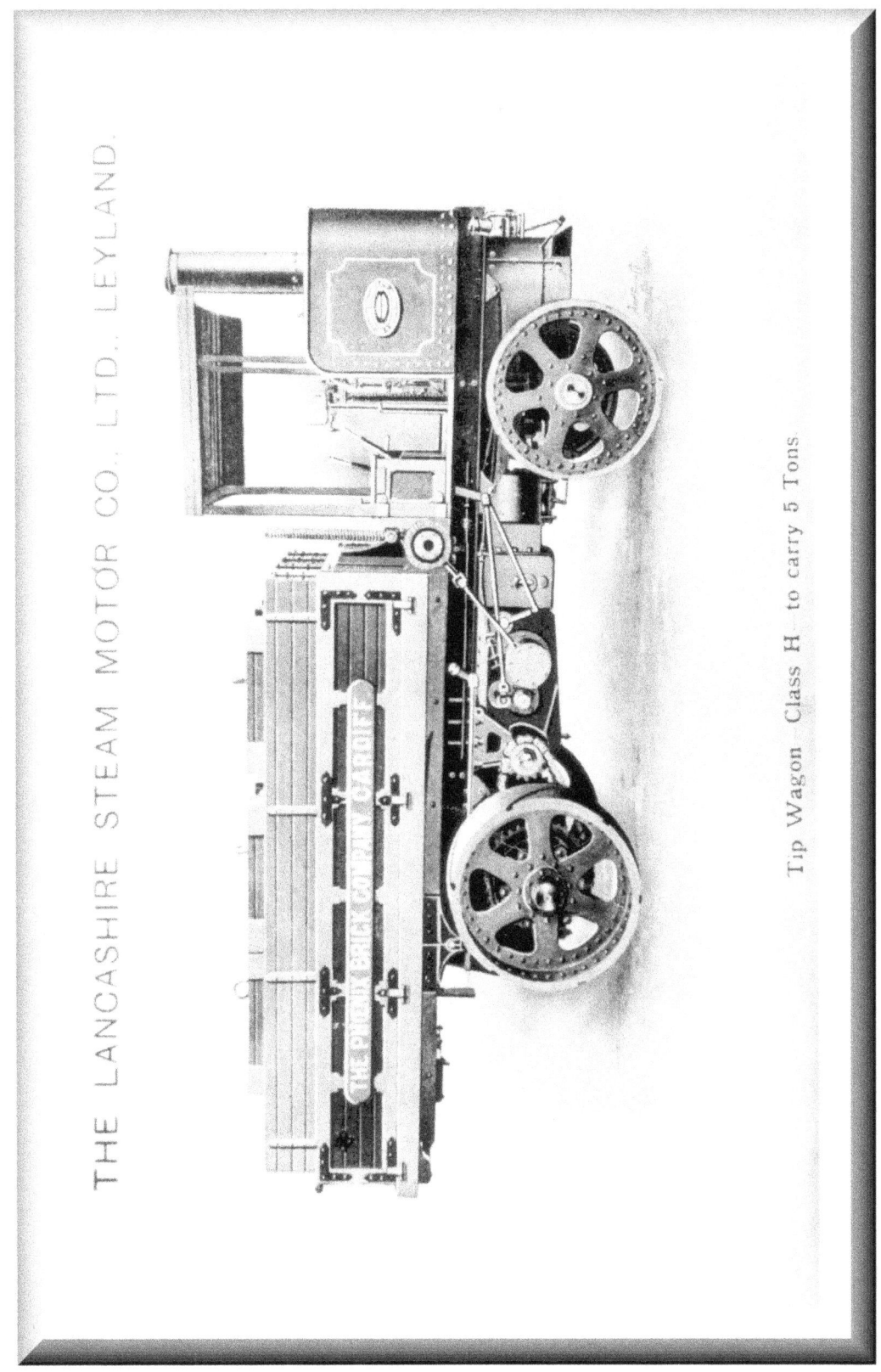

Tip Wagon – Class H – to carry 5 Tons

THE LANCASHIRE STEAM MOTOR CO., LTD., LEYLAND.

Kevin Lomas

Kevin Lomas

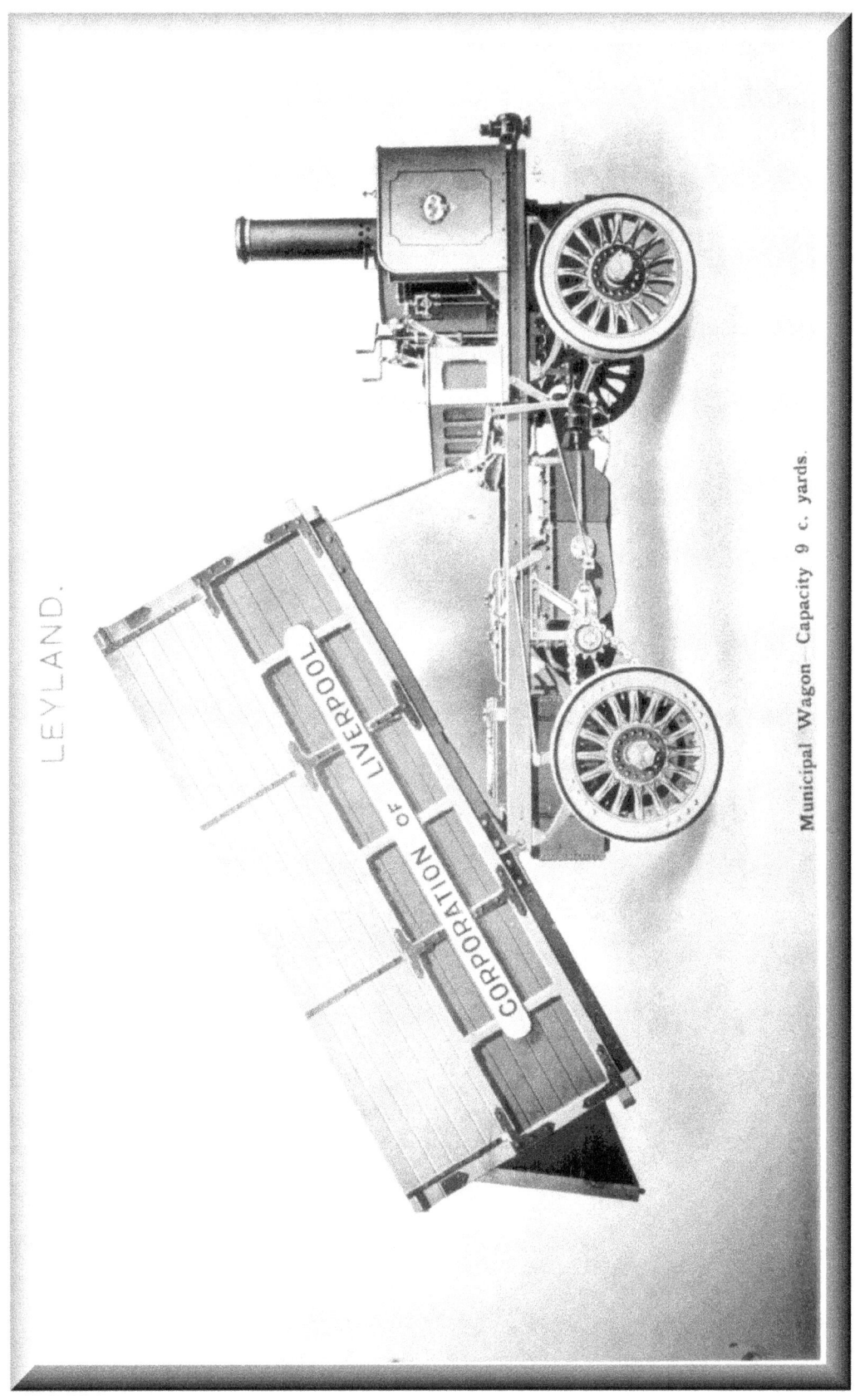

LEYLAND.

Municipal Wagon—Capacity 9 c. yards.

Kevin Lomas

"Colonial" or Contractors Tip Wagon—4, 6, and 8 Tons.

Steam Vehicles Of The Road

Kevin Lomas

Other Works By Kevin Lomas –

Science/Fantasy Fiction –

Lilium Saffron Dewbell – Part One – Once Upon a Time – Leaving the heavily protected dimension of Fairyland, to go about her traditional task on Terra, something terrible and unheard of occurs to our sweet, innocent and 'young' Fairy Lilium, and that is only the start of her many problems as she later discovers that nothing she has taken for granted is as it seems and she becomes far less sweet and innocent.

Lilium Saffron Dewbell – Part Two – Out There - … and Lilium, disgusted with her gods, at last ventures out into the universe/s and dimensions on her own. But what is the point of being told you are becoming a goddess, and potentially dangerous, when even your natural Fairy magic is often on the blink? And still she has not had her vengeance …

Lilium Saffron Dewbell – Part 3 – You Cannot Bottle a Goddess – Lilium continues her amazing and dangerous adventures and soon discovers that friends are not always friends because she is a valuable commodity. She also has a good lead on the evil Dark-Fairy who started all of this. Shame that that planet is currently at war.

Short Stories –

Demonstrative – It started on a nice day, the happiest of most people's lives. And then it turned bad. Then it got worse. No one had any idea how bad it could get. Religion and cults warned us, and now it is happening. 'Evil' from the depths of time is back, and it is not happy …

Project Thirian – Making a nearby planet habitable is cause for celebration, but trillion to one chances happen too often.

Just a Couch – A Long Poem – All they wanted was just a couch …

Circus – A circus is coming to town, but this is unlike any other circus …

All available printed or as e-books.

Art –

Plus a vast amount of original printed artwork of various types available in many sizes, on from paper to canvas, framed or not framed.

All viewable and available for purchase via –
kevinlomas.net

www.ingramcontent.com/pod-product-compliance
Lightning Source LLC
Chambersburg PA
CBHW081049170526
45158CB00006B/1914